SLOPES, ANGLES, AND RATIOS

ROADS AND RAMPS

MathScape
SEEING AND THINKING
MATHEMATICALLY

In this phase, you will focus on tools used to measure the slope angles of slanted objects. You will make your own slope-measuring device and use it to measure slopes inside and outside your classroom. You will also make scale drawings of the objects you measure.

What math is used in designing and building roads and ramps?

ROADS AND RAMPS

PHASE**TWO**
Right-Triangle Relationships

You will measure and calculate the side lengths and slope angles of right triangles. As you study the side length relationships of right triangles, you will compare two different methods for calculating side lengths—scale drawings and the Pythagorean Theorem.

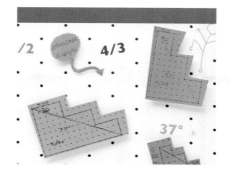

PHASE**THREE**
Slope as a Ratio

As you measure the vertical rise and horizontal run of some stairs and ramps, you will look for relationships between slope ratio, slope angle, and percent grade. To wrap up the phase, you will have a chance to design a set of stairs.

PHASE**FOUR**
Road Design

In this final phase, you will create two- and three-dimensional road models. You will learn about a helpful mathematical tool—the tangent ratio—and then build a hillside model that you can analyze and measure. Then it's your turn to design and build your own model.

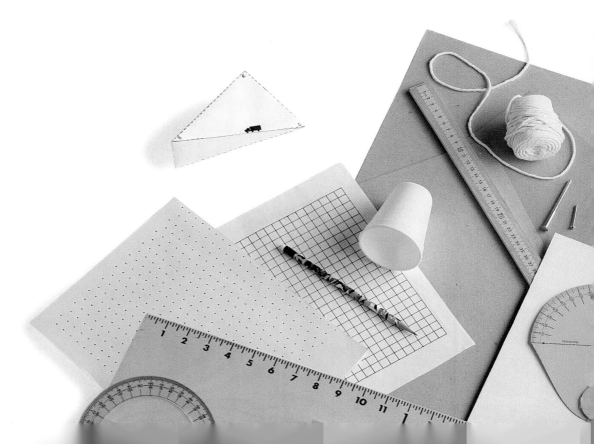

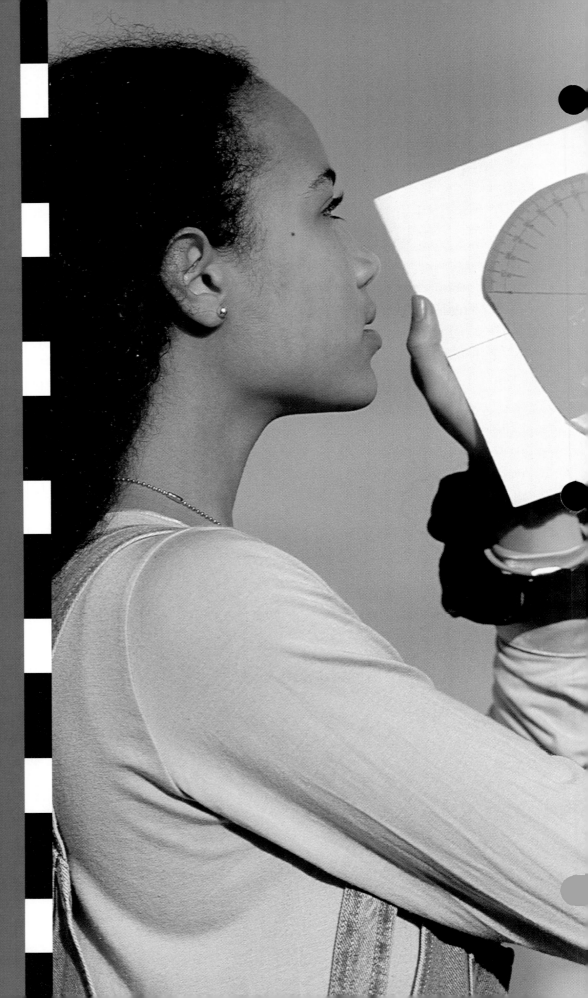

PHASE ONE

In this phase, you will measure the slope angles of slanted objects using a Slope-o-meter™ and a protractor.

Every profession has specific tools that are used by the people in that field. The Slope-o-meter that you will work with is similar to a protractor, a tool used in many occupations. Besides math students, who else do you think uses a protractor?

Slope as an Angle

WHAT'S THE MATH?

Investigations in this section focus on:

SCALE and PROPORTION

- Recognizing the difference between a sketch and a scale drawing.

- Relating slope angles sighted with the Slope-o-meter to angles in scale drawings that are drawn with a protractor.

- Using ratios to relate lengths in scale drawings to actual lengths.

ACCURACY and MEASUREMENT

- Estimating and measuring the slopes of lines as angles (from the horizontal) to the nearest degree.

- Recognizing horizontal and vertical benchmarks for slope.

- Examining the relationships among the accuracy of measurements, the quality of the measuring tool, and the limitations of human perception.

MathScape Online
mathscape3.com/self_check_quiz

1 Slopes and Slope-o-meters

MEASURING SLOPE AS AN ANGLE

Most roofs, roads, and ramps in our world have a slope. To begin your investigation of these, you will make and use a tool for measuring slopes.

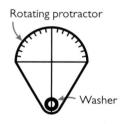

Making a Slope-o-meter

To make a Slope-o-meter, you will need two pieces of printed cardstock. One piece has the rotating protractor, and the other has the backing for the Slope-o-meter. You will also need a heavy washer and a pierced-earring post and back.

1. Cut out the rotating protractor and tape a heavy washer to the bottom of it. Be sure to center the washer over the vertical line, so that this part of the Slope-o-meter will hang vertically.

Rotating protractor

Washer

2. On the other piece of cardstock, score the dotted lines with a pen or paper clip. Then fold back the edges along the dotted lines and tape them to the back of the cardstock. This will make the backing of your Slope-o-meter sturdy.

Backing

3. Use transparent tape to reinforce the swivel point of the rotating protractor with an "X." Do the same at the center of the backing.

Tape

Tape

4. Use the earring post to pin the swivel point of the rotating protractor to the center of the backing. Your completed Slope-o-meter should look like this.

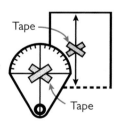

Back

Front

Earring back

Earring post

Measure Slopes

Use your Slope-o-meter to measure the slopes of lines posted around the classroom.

1 Before you measure each slope, make an estimate and record it in a table.

2 After you make an estimate, use the Slope-o-meter to measure the slope of the line. Record the measurement in your table. As you record the slopes, think about the accuracy of your Slope-o-meter and consider reasons why different students might find different slopes for the same line.

3 How do your estimates and actual measurements compare?

4 How do your measurements compare with those of your classmates?

How can you use your Slope-o-meter to measure slopes?

How to Use a Slope-o-meter

You can rest your Slope-o-meter directly on a surface to measure its slope. You can also align the bottom edge of the Slope-o-meter with a line drawn on the chalkboard.

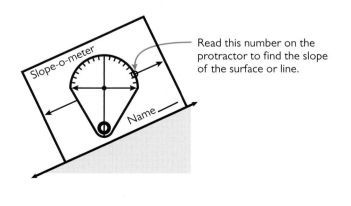

Read this number on the protractor to find the slope of the surface or line.

hot **words** | slope angle

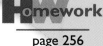

page 256

2 Working for Scale

If you were going to construct a ramp or road, you might begin by making an accurate scale drawing. You will get a chance to make your own scale drawings after first using a protractor to measure some slopes.

Measure Slopes with a Protractor

How can a protractor help you measure slopes?

Consider the slopes in these pictures below and then record this information in your lab journal.

1 Order these slopes from least to greatest.

2 Estimate each slope in degrees measured from the horizontal.

3 Use a protractor to measure the slope in each picture.

4 Make a table comparing your estimated slope measurements to the actual slopes.

Ladder

Roof

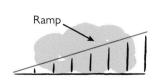

Ramp

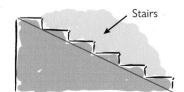

Stairs

Read a Scale Drawing

A **scale drawing** is an accurate drawing of a life-size object that is proportionally smaller. Architects, physicists, and engineers use such drawings to help study specific parts of a situation or problem.

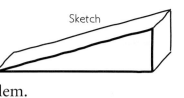
Sketch

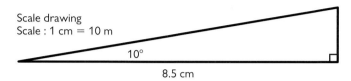
Scale drawing
Scale : 1 cm = 10 m
10°
8.5 cm

Use the scale drawing shown here, as well as a protractor and ruler, to answer the following.

1 How tall is the drawing of the ramp in centimeters?

2 What is the height of the actual ramp in meters?

3 What is the measure of the third angle in the drawing? How do you know you are correct?

4 How would the slope of the ramp change if its height increased? decreased?

What information can you get from a scale drawing?

Make a Scale Drawing

Choose a scale and make an accurate scale drawing using a ruler and protractor to solve the following problem.

- A ramp with a 20° slope is 32 ft long. What horizontal distance does the ramp cover? What vertical distance does it rise?

How can scale drawings be used to find unknown lengths?

hot **words** | scale
scale drawing

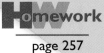
Homework

page 257

3 Sighting and Angle of Elevation

USING A LINE OF SIGHT IN SCALE DRAWINGS

You can use your Slope-o-meter to measure the slopes of inaccessible objects, such as a faraway roof or a mountain, with a method called sighting. Your Slope-o-meter can also be used to find the slope of your line of sight to an object.

Use a Slope-o-meter for Sighting

How can you use sighting to find the slope of a distant object?

Use sighting to measure the slope of the roof line that your teacher has drawn. Sketch the slope and record the measurement in your lab journal.

As you work, think about sources of error and why different students might get different readings.

The Sighting Method

To find the slope of an object by sighting, hold out your Slope-o-meter. Line up the bottom edge of the Slope-o-meter with the slope line you are measuring.

Slope-o-meter

Make a Scale Drawing of a Flagpole

Measure 20 m out from the base of a flagpole at school. Stand at that point and use your Slope-o-meter to find the angle of elevation from your eye level to the top of the flagpole. Then create a scale drawing of the situation in your lab journal. Don't forget to include the distance from the Slope-o-meter to the ground in your scale drawing.

1 What is the height of the flagpole?

2 How does your result compare with those of other students?

3 How accurate do you think your result is? Is it accurate to the nearest 1 cm? the nearest 5 mm?

> **How can you use an angle of elevation and a scale drawing to find the height of a flagpole?**

How to Measure an Angle of Elevation

You can use your Slope-o-meter to find an angle of elevation. The angle of elevation is the angle up to an object, measured from the horizontal.

To measure an angle of elevation, you must sight along the top (or bottom) edge of your Slope-o-meter, holding the Slope-o-meter close to your eye.

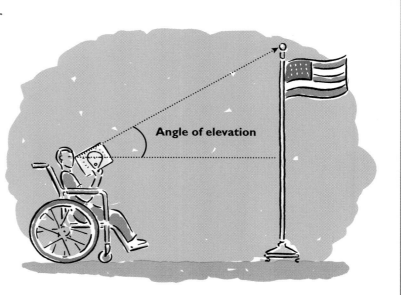

Angle of elevation

hot **words** | angle of elevation

Homework
page 258

PHASE TWO

In this phase, you will find the lengths and slope angles of some right triangles using scale drawings and the Pythagorean Theorem.

Understanding scale drawings and the Pythagorean Theorem gives you more than one way to find the unknown lengths. Being able to solve problems in different ways is an important skill. How might an architect or a surgeon solve problems in more than one way?

Right-Triangle Relationships

WHAT'S THE MATH?

Investigations in this section focus on:

GEOMETRY

- Visualizing horizontal, vertical, and slope lines as right triangles.

- Understanding the relationship between the acute angles of any right triangle.

- Using vocabulary associated with right triangles.

- Applying the Pythagorean Theorem in problem-solving situations.

ACCURACY and CALCULATIONS

- Making scale drawings to improve accuracy.

- Using the Pythagorean Theorem to calculate unknown side lengths in right triangles.

- Comparing measurements found with scale drawings to those calculated with the Pythagorean Theorem.

MathScape Online
mathscape3.com/self_check_quiz

4 Right Triangles

EXPLORING ANGLE RELATIONSHIPS IN RIGHT TRIANGLES

Imagine you are climbing up a ladder that is leaning against a building. Is the ladder leaning at a stable and safe angle? In this lesson, you will experiment with the slope angles of ladders to learn about angle relationships in right triangles.

Explore Ladder Stability

How can you figure out the maximum and minimum safe slopes for a ladder?

Set up the experiment you see here.

1. Tape a paper cup to a meterstick. The cup should be placed near the top, as this is the critical position when a person is on a ladder.

2. To simulate the weight of a person on the ladder, put something into the cup such as wooden blocks or keys.

3. Lean the stick against a wall.

Record your work as you investigate the following:

- Find the maximum and minimum stable slopes for the ladder.

- Use a Slope-o-meter to measure each of these slope angles.

- For each slope, use a protractor to measure the angles the ladder makes with the floor and with the wall.

- Which of the protractor measures is the same as your Slope-o-meter measure? Why?

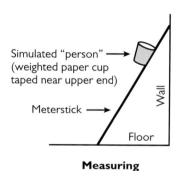

Simulated "person" (weighted paper cup taped near upper end)

Meterstick →

Wall

Floor

Measuring with Slope-o-meter

Measuring with protractors

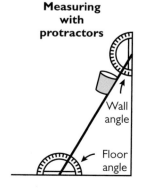

Wall angle

Floor angle

Find Ladder Angles

Find the wall angle for each of the ladder's floor angles below. You may want to use a protractor or Slope-o-meter first to help position the meterstick so that it forms each angle with the floor.

1. 30°   2. 50°   3. 70°   4. 63°   5. $x°$

What is the relationship between floor angles and wall angles?

Analyze Right Triangles

The ladder experiments that you have done can be represented by drawing right triangles. These diagrams can help you to better understand the relationship between horizontal, vertical, and sloped lines.

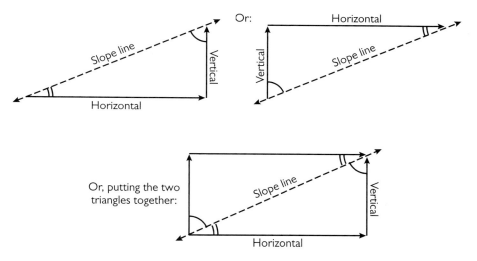

Use a protractor and ruler to draw these diagrams in your lab journal. Then write answers to the following questions.

1 How do these diagrams help explain the angle relationships in the ladder experiments you just did?

2 What can you say about how the acute angles in a right triangle are related?

3 Is this relationship true for triangles that are not right triangles? Why or why not?

hot **words** | right angle
right triangle

Homework
page 259

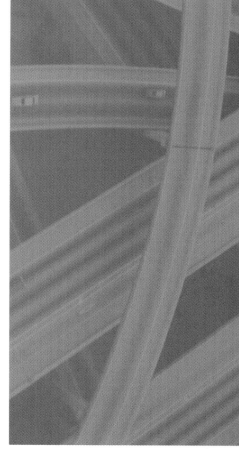

5 Exploring the Pythagorean Theorem

EXPLORING SIDE
RELATIONSHIPS IN
RIGHT TRIANGLES

As you've seen, a ladder leaning against a building forms a right triangle. Thousands of years ago, people discovered something very special and remarkable about the relationships among the sides of a right triangle. Here you will create some right triangles and explore the relationships among the sides of each triangle.

Measure the Sides of Right Triangles

What relationship can you find among the sides of a right triangle?

Follow the steps below to make right triangles, measure the lengths of the sides, and record your data in a table.

1 Cut out four different L-shaped strips of centimeter graph paper. Each strip should be one centimeter wide. A sample strip is shown here.

2 Use a metric ruler to measure across the inside of each L-shaped strip. Measure this distance to the nearest $\frac{1}{10}$ cm. Record this length and the inner leg lengths in a table.

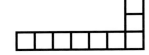

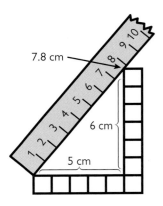

7.8 cm

6 cm

5 cm

3 Draw and label the right triangle that corresponds to each L-shaped strip.

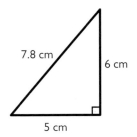

7.8 cm

6 cm

5 cm

Write About Right Triangles

Discuss each of these questions with a partner and then record your answers in your lab journal.

1 The *angles* of the right triangle have a specific relationship. Describe that relationship.

2 Write a general rule that describes the relationship among the *sides* of a right triangle. Include a drawing with your rule.

Explore Pythagorean Triples

The whole numbers 3, 4, and 5 are known as a **Pythagorean triple.**

1 Write the Pythagorean Theorem equation that corresponds to the triangle shown.

2 What do you think is meant by the term *Pythagorean triple*?

3 The numbers 6 and 8 are two members of another Pythagorean triple. That is, $6^2 + 8^2 = c^2$.

   a. Solve for c.

   b. Compare the side lengths of this triangle with those of the 3-4-5 triangle.

4 Find the missing values in each set of Pythagorean triples.

   a. 30, 40, ?

   b. 15, ?, 25

   c. ?, 44, 55

5 Given one Pythagorean triple, how can you find new ones?

   a. Here is another Pythagorean triple: 5-12-13. Verify that this is a Pythagorean triple.

   b. Use what you have discovered above to find some new Pythagorean triples.

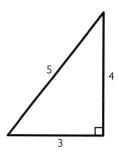

What can you say about whole numbers that satisfy the Pythagorean Theorem?

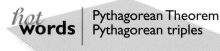

hot **words** | Pythagorean Theorem
| Pythagorean triples

Homework

———————
page 260

6 Using the Pythagorean Theorem

Buildings are designed and constructed with a specific degree of accuracy. Boards or beams that are an inch too short or too long can cause major problems in construction. Here you will use the Pythagorean Theorem to help find the length of a beam for a building.

Compare Scale Drawings with the Pythagorean Theorem

How can you accurately determine the length of a ramp?

Suppose you are going to make a ramp for your school. It will start at the sidewalk and end at the top of a stairway that leads to the front of the school. The ramp is to cover a horizontal distance of 40 ft and a vertical distance of 5 ft.

5 ft

40 ft

1 Choose a scale and make an accurate scale drawing using a ruler and protractor to solve this ramp problem. What is the length of the ramp?

2 Use the Pythagorean Theorem to solve the same problem.

3 How do your two results compare? How do the two methods compare? Do you think one of the methods is more accurate? Explain your thinking.

4 How long would a banister for the ramp be if it needs to extend past each end of the ramp by 1 ft?

Apply the Pythagorean Theorem

Imagine that you work for Envirotec Builders. Your boss asks you to order a set of steel beams that will support the sloped roof of a building and overhang each end by 1 ft. The length of the beams must be very accurate because there is no way to cut the beams on-site after they arrive.

Here is a side view of the building (not drawn to scale).

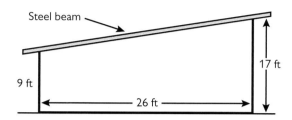

- How long should the beams be? (Remember, the steel beams must extend past each wall by 1 ft.) Show all of your work.

- Describe in writing the method you used to find the length of the beams. Did you make a scale drawing or use the Pythagorean Theorem? How accurate do you think your answer is?

The Pythagorean Theorem

For any right triangle, the sum of the squares of the legs equals the square of the hypotenuse.

$$a^2 + b^2 = c^2$$

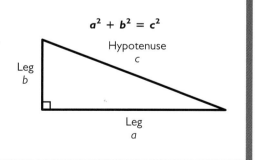

hot **words** | scale drawing
Pythagorean Theorem

Homework
page 261

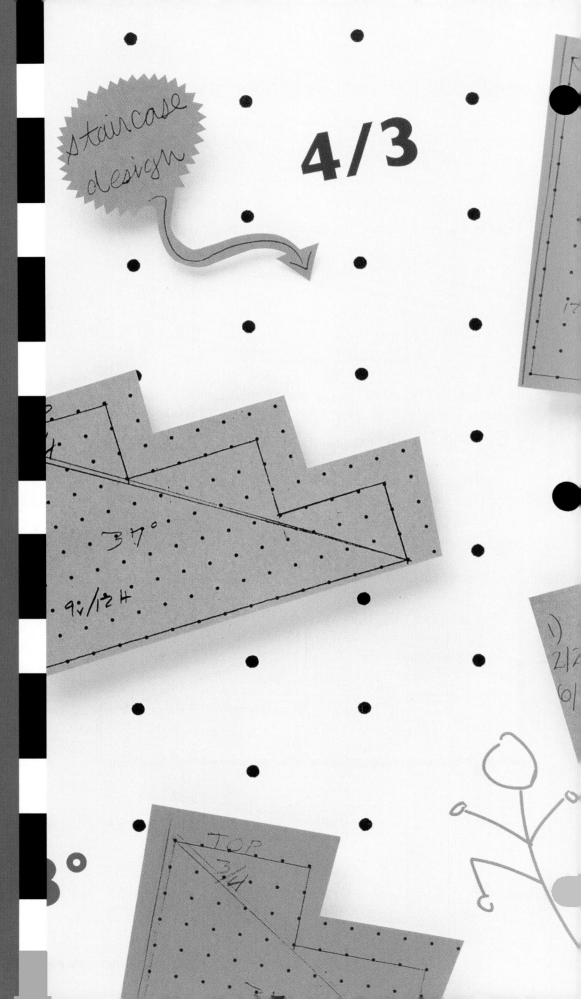

PHASE THREE

4/3

staircase design

TOP
3/4

In this phase, you will find the slope angles of stairs using ratios. You will have a chance to design your own staircase. You will also do some experiments to investigate the relationship between slope angle, slope ratio, and percent grade.

Understanding how to use ratios is very important in many occupations. Architects use ratios to help make models of buildings they are designing. What other professions do you think use ratios? How do you think they use them?

Slope as a Ratio

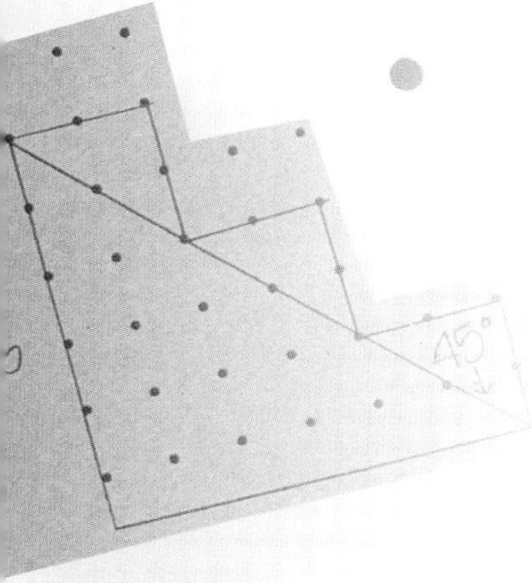

WHAT'S THE MATH?

Investigations in this section focus on:

RATIOS and MEASUREMENT

- Understanding the definition of slope as a ratio of vertical rise to horizontal run.

- Calculating the slope ratios of stairs and other objects.

- Understanding how percent grade is used to describe slope ratios.

- Finding the slope angle for a given percent grade.

PROPORTIONAL REASONING

- Setting up proportions to solve real-world problems.

- Solving proportions to find equivalent ratios.

MathScape Online
mathscape3.com/self_check_quiz

7 Stairs and Ratios

You've already seen how to measure a slope as an angle in Phase One of this unit. In these next few lessons you will explore how carpenters, scientists, and mathematicians measure slope as a ratio.

Find the Slope Ratio of Stairs

How can you determine the slope ratios for different sets of stairs?

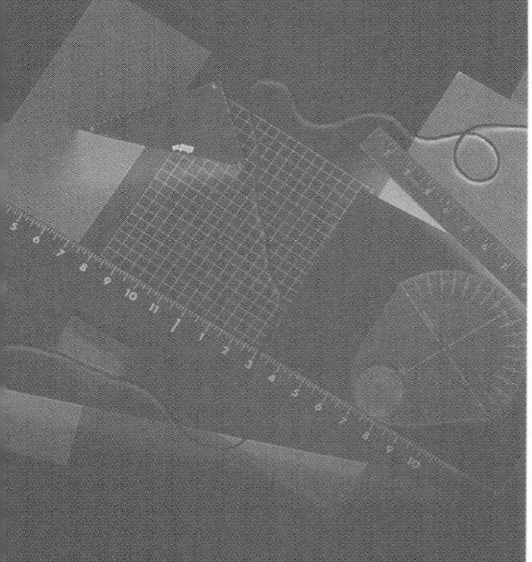

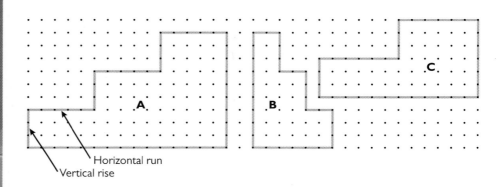

Horizontal run

Vertical rise

$$\text{Slope ratio} = \frac{\text{Vertical rise}}{\text{Horizontal run}}$$

Answer these questions for each set of stairs, A, B, and C. Record your findings in a table like the one shown.

1 What is the slope ratio?

2 Estimate the slope angle.

3 What is the actual slope angle? Measure it with a protractor.

Stairway	Slope Ratio	Slope Angle	
		Estimate	Actual
A			
B			
C			

Design Steps for Different Slope Ratios and Angles

Use centimeter dot paper to help answer the following questions.

1 Draw a three-step stairway for each of the following slope ratios.

   a. $\frac{2}{5}$ **b.** $\frac{5}{2}$ **c.** $\frac{3}{4}$ **d.** $\frac{4}{3}$ **e.** $\frac{5}{8}$

   f. $\frac{8}{5}$ **g.** $\frac{5}{6}$ **h.** $\frac{6}{5}$ **i.** $\frac{2}{2}$ **j.** $\frac{3}{3}$

2 What is the slope angle of each of the stairways you drew in item **1**? Measure your drawings with a protractor.

3 **a.** For each set of stairs you drew in item **1**, find the total vertical rise and total horizontal run.

   b. In each case, how are the total rise and total run for the three steps related to the rise and run for one step?

4 If more steps with the same slope ratio are added to a stairway, will the slope angle of the stairway change? Draw a sketch to explain your answer.

5 Building codes say that handicapped-access ramps can have a slope ratio no greater than $\frac{1}{12}$. Which of these slopes would be okay? Why?

   a. $\frac{3}{35}$ **b.** $\frac{4}{50}$ **c.** 20°   **d.** 10°

6 Do you think the following statement is true?

If you have two slope ratios that are reciprocals, then the corresponding slope angles for those ratios will add up to 90°.

Test the statement on the stairs with reciprocal slope ratios that appear in item **1**, such as $\frac{2}{5}$ and $\frac{5}{2}$. Can you explain why the statement might or might not be true?

How can you draw stairways with different slope ratios?

hot **words** | slope ratio

Homework

page 262

ROADS AND RAMPS • LESSON 7   **243**

8 The 12-Inch Tread

Most lumber yards sell precut boards called 12-inch steps, which are used for making stair treads. The boards are Douglas fir and are 1 in. thick, 12 in. wide, and can be cut in various lengths. In this lesson, you will investigate how to use 12-in.-wide boards to create some specific stairways.

Explore Stairs with a 12-Inch Tread

How can you determine the rise of a step, given the tread and slope of a staircase?

A carpenter needs your help in making a stairway. Each step is to have a 12-in. horizontal tread. The slope of the stairway is to be $\frac{2}{5}$.

- Given the 12-in. tread, what should the rise for each step be? Work with classmates to solve this problem in at least two different ways.

- Record your work in your lab journal and be prepared to share your methods with the class.

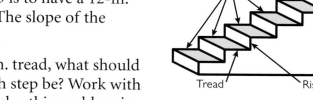

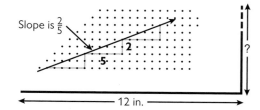

Share and Compare Solution Methods

Present your findings to the whole class.

- Take notes in your lab journal while other groups are making their presentations.

- After all groups have shared their findings, choose a method you can use quickly to calculate the rise for a 12-in. tread for any slope ratio. Explain how the method works.

Design a Stairway

The guidelines shown are from a stair-building manual. Imagine that you are a master carpenter and need to design a stairway that goes from the basement to the first floor of a house. The total rise of the stairway must be exactly 10 ft and must follow the guidelines. Your stairway may have a different number of steps from the one shown in the diagram, and it is not necessary for each step to have a 12-in. tread.

Write up your design recommendations for the stairway. Be sure to show your work and include the following information.

1 What are the maximum and minimum possible values for the total run of the stairway?

2 What is your recommended design for the stairway?

  a. How many steps would you have?

  b. What is the total run of your recommended staircase?

  c. What is the unit run?

  d. What is the unit rise?

How can you design a stairway, given guidelines for the rise and run?

Stair-Building Design Guidelines

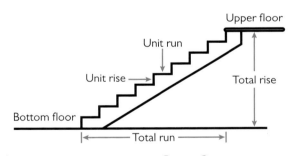

Step 1

Calculate unit rise by first choosing the number of rises for your design. When divided into total rise, the result should be a unit rise of 6–8 in. (As an example, the diagram above shows 9 rises.)

Step 2

For comfortable stairs, calculate unit run with this formula:

unit rise + unit run = between 17 in. and 18 in.

Step 3

Add up the unit runs to find total run.

hot **words** | tread slope

Homework

page 263

9 7% Grade Ahead!

CONNECTING SLOPE
RATIOS AND
PERCENT GRADE

The 7% Grade Ahead sign is a truck warning sign at the start of a steep hill on Highway 92 in northern California. In this lesson, you will explore percent grades.

Experiment with Slope Angles and Percent Grades

What is the relationship between slope angles and percent grade?

Suppose you work for a survey crew that needs a chart that relates the slope angle for a hillside to its percent grade. Here is a sketch that shows how you can get the data for the chart.

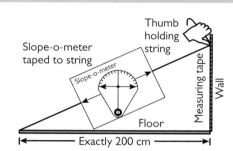

1. Your teacher will hand out an Angle and Percent Grade Chart. Leave the far right column blank for now—it will be used in a later lesson.

2. Tape the backing of your Slope-o-meter to a piece of string, as shown in the figure.

3. Hold or tape one end of the Slope-o-meter/string device to the floor so that it is 200 cm away from a wall.

4. Hold the other end of the string against the wall and move it up or down until the Slope-o-meter shows the desired slope angle.

5. Use the tape measure to find the resulting wall height. Record the wall height in your chart. This tells you how high a hillside with that slope angle rises for every 200 cm of horizontal distance.

6. Find the slope ratio and convert it to a percent grade. Record the value in your chart.

Percent Grades

A percent grade of 7% means that for every 100 units of horizontal distance traveled, the road ascends (or descends) 7 units.

7% Grade Ahead

Design a Skateboard Ramp

How can you design a ramp with a specific percent grade?

Ramps-R-Us makes skateboard ramps at whatever percent grade the customer orders. The boss wants you to help design the vertical braces for the ramps.

"All our ramps have the same horizontal length (18 ft) and six braces, each 3 ft apart," she says. "I'm giving each of you a different percent grade to work with. Figure out the heights of all the braces and the total length of a ramp with that percent grade. Then when we get an order, all we will have to do is pull up the specs for that percent grade and build the ramp."

Prepare a report for your boss that includes the following.

1 A labeled diagram of your ramp that shows:

   a. the percent grade of your ramp

   b. the height of each of the six braces

   c. the total length of the ramp

2 A description of how you figured out the height of the braces and the total length of the ramp.

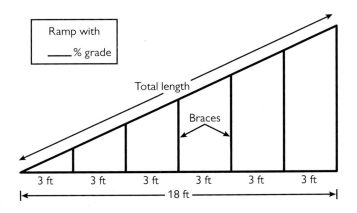

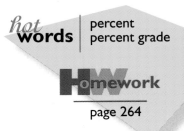

hot **words** | percent
percent grade

Homework

page 264

PHASE FOUR

In this phase, you will explore the tangent ratio. The tangent ratio ties together the ideas of slope as an angle and slope as a ratio. Then you will construct a three-dimensional model of a road on a hill.

People solve all types of problems in art, math, science, and engineering by using models. What are some advantages of making and analyzing models?

Road Design

WHAT'S THE MATH?

Investigations in this section focus on:

TRIGONOMETRY

- Finding the tangent ratio for a given slope angle.

- Finding the slope angle that corresponds to a given tangent ratio.

- Calculating unknown side lengths and angle measures in right triangles.

SPATIAL RELATIONSHIPS

- Visualizing the spatial relationships among horizontal, vertical, and slanted planes.

- Relating slopes in three dimensions to their two-dimensional representations.

SCALE MODELS

- Indentifying and calculating the essential lengths and angle measurements needed to construct a three-dimensional model.

- Devising a strategy for calculating unknown side lengths and angle measures.

- Building an accurate three-dimensional scale model.

MathScape Online
mathscape3.com/self_check_quiz

10 The Tangent Ratio

LINKING SLOPE RATIOS AND SLOPE ANGLES

In previous lessons, you investigated slope angle, slope ratio, and percent grade. In this lesson, you will investigate the tangent ratio and see how it ties all of these ideas together. This will help you fill in the last column of your chart from Lesson 9.

Calculate the Tangent Ratio

How are slope angles and slope ratios related?

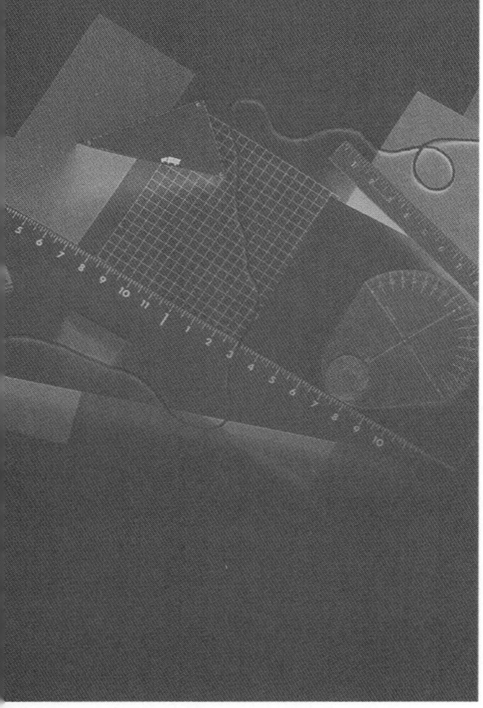

Here is a way to explore the relationship between slope angles and slope ratios.

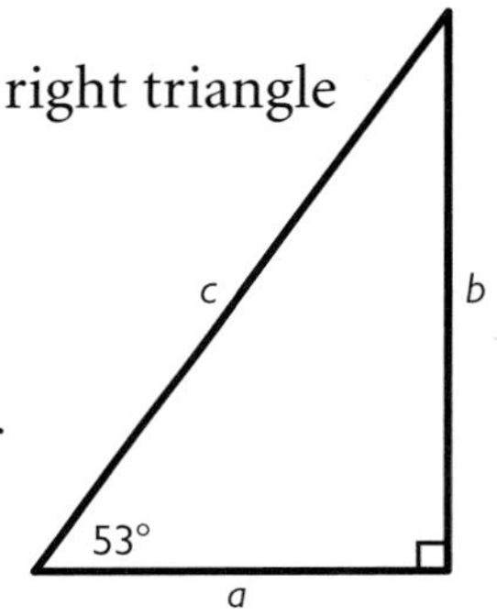

1. Each member of your group should draw a right triangle (of any size) with a slope angle of 53°.
2. Use a ruler to find the slope ratio of your triangle and convert it to a decimal.
3. Compare your decimal value with the other decimal values in your group.
4. What do you notice about each value?

The ratio you calculated is known as the tangent ratio for 53°. The tangent ratio for a given slope angle is constant, no matter how large or small the right triangle is that contains it.

The **tangent** of an acute angle in a right triangle is the ratio of the length of the opposite side to the length of the adjacent side (or $\frac{b}{a}$ in the figure).

Find Angles, Grades, and Rises

Create a table that shows percent grades from 0% to 100% in increments of 10%. Include a column for rises for a set of stairs with a 12-in. tread. Here is an example. Only the first few rows are shown, with one row completely filled in as an example.

Slope Angle	Slope Ratio as Decimal (tangent ratio)	Slope Ratio as Percentage (% grade)	Rise for 100 Feet of Run	Rise for 12-Inch Tread
		0%		
		10%		
		20%		
16.7°	0.3	30%	30 ft	3.6 in.

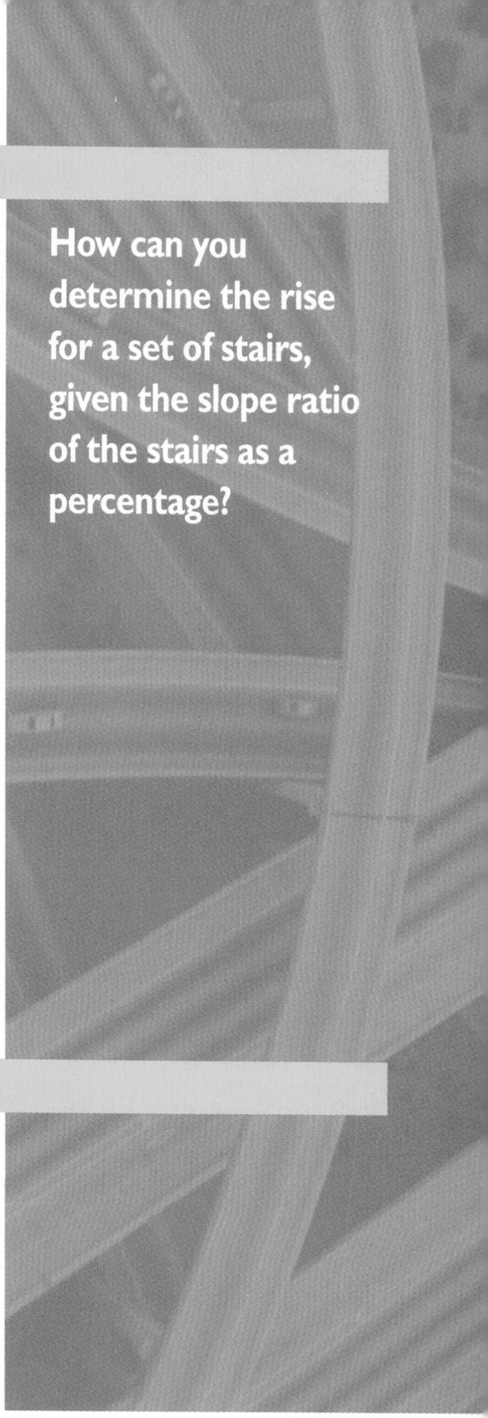

How can you determine the rise for a set of stairs, given the slope ratio of the stairs as a percentage?

1 What is the rise for a 12-in. tread with a slope angle of 60°? 75°?

2 What is the percent grade of stairs with a rise of 8.5 in.?

3 What are some easy ways to find the rises for a 12-in. tread?

Solve Right Triangle Problems

Your goal is to find all the unknown side lengths and angles for the triangles shown below and write an explanation of your work. Note that the triangles are not drawn to scale!

1 Use the tangent ratio, the Pythagorean Theorem, and any other methods you know to find all of the unknown sides and angles for each triangle.

2 Write a step-by-step explanation of how you solved these problems.

3 Have a classmate look at your explanations. Can he or she use your method to solve other such right triangle problems?

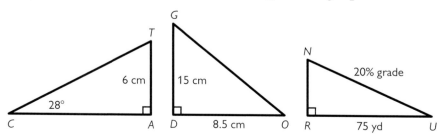

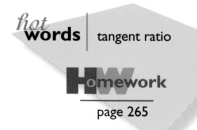

hot **words** | tangent ratio

Homework

page 265

11 A Mathematical Hill

When building a road on a steep hill, most engineers design the road to be less steep than if it went straight up the hill. In this lesson, you will make a three-dimensional model of a hill and analyze different roads that travel up it.

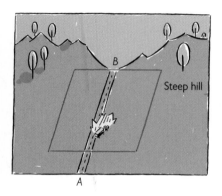

We will "cut out" the rectangular portion of the hillside shown and look at it more closely:

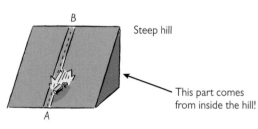

This part comes from inside the hill!

Explore a Road Model

How can you compare different roads that go up a steep hill?

An engineer looked at Road Design 1 and decided it was too steep for a truck to go straight up along line *AB*. The engineer began working on a new road design for the hillside and came up with Road Design 2.

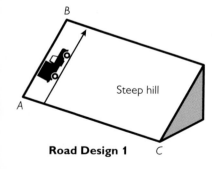

Road Design 1

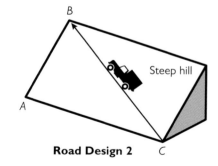

Road Design 2

1 How does line *AB* of Road 1 compare to line *CB* of Road 2?

2 Why is the slope of Road 2 a better design than that of Road 1?

3 If you were asked to make a three-dimensional scale model of this situation, what measurements would you need?

Analyze the Hillside

Use your model and a Slope-o-meter to analyze the hillside.

How can you use a model to analyze a steep hill?

1 Measure all of the hillside model's angles and lengths with a protractor and metric ruler. Record the measurements for each triangle on the Model Hillside Recording Sheet.

2 What do you think is the steepest road up the model hill? Use your Slope-o-meter to find and measure the slope angle of that road. Which of the slope angles that you measured with a protractor corresponds to the slope angle you just measured with the Slope-o-meter? How close are the two measurements?

3 Use your Slope-o-meter to measure the slope angle of the road going from *C* to *B*. Which slope angle measured with a protractor matches this slope angle? How close are the two?

4 For road *AB*, what is the vertical rise? the horizontal run? the percent grade?

5 For road *CB*, what is the vertical rise? the horizontal run? the percent grade?

6 What are some reasons that a truck might use road *CB* instead of road *AB*?

How to Assemble the Model Hillside

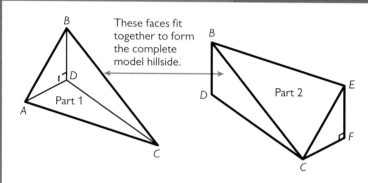

These faces fit together to form the complete model hillside.

Work with a partner to make a model of a hillside, as shown in the diagram. This will help you see the mathematics underlying a road design.

1. Carefully cut out Model Hillside Part 1 and Model Hillside Part 2.

2. Use a ballpoint pen and ruler to score the dotted lines. This will help make the folds clean and precise.

3. Fold along the dotted lines. As you fold, be sure the printed lines of each part are on the outside of the model. Tape the edges of each part of the model together.

4. Place the two pieces together on a flat surface.

hot **words** | steepness slope angle

Home**work**

page 266

12 The Road Project

CREATING A SCALE DRAWING AND A 3-D MODEL

In this lesson, you will be given information about a new hill model. It will be up to you to figure out the size and shape of the triangles that make up the model.

Explore a Sample Road Model

How can you find the unknown lengths and angles of the four triangles in this model?

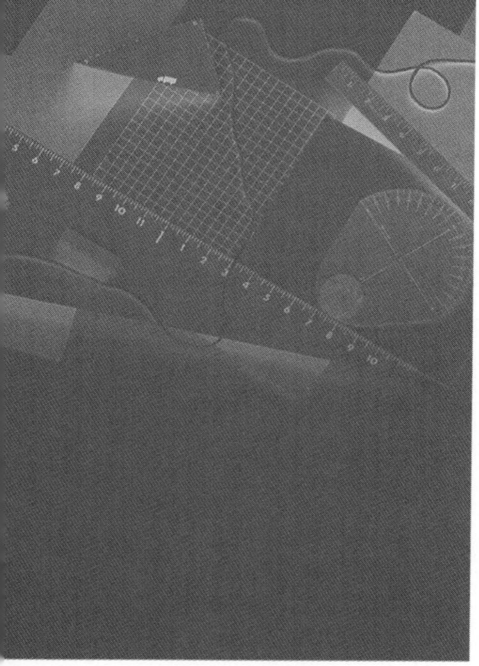

A hillside has a slope angle of 38° and a vertical rise of 800 ft. A road with a 30% grade is to go up this hill. Your design team needs to make scale drawings of the four triangles that make up the model (shown here with heavy lines).

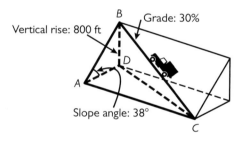

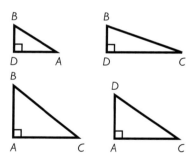

Here are some steps to follow to help you and your team make a scale drawing of the four triangles in this model. (The triangles shown here are not to scale, but you may want to copy them into your lab journal and record your results on them.)

1 To build a model for this hill, do you need to find *all* the lengths and angles? Record the known lengths and angles. Then make a list of lengths and angles that you still need to find.

2 Why must a length in one triangle match a length in another triangle? For example, why must $\overline{BC}$ in $\triangle ABC$ be the same length as $\overline{BC}$ in $\triangle DBC$?

3 Find the unknown lengths and angles.

4 Use a ruler and protractor to make a scale drawing of each of the triangles.

Calculate the Dimensions for the Model

Your teacher will give you the following three starting measurements for your model.

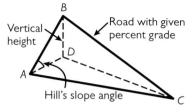

Angle	the slope angle of the hillside
BD	the vertical rise of the hill in feet
Percent grade	the percent grade of the road *BC*

Use this information to figure out the unknown lengths and angles of each triangle.

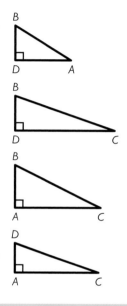

1 Record the starting measurements on the Road Model Record and Evaluation Sheet.

2 On the back of the project sheet, draw and label the four right triangles you see here. (Note: The drawings are not to scale.)

3 Find the angles and lengths needed to draw each triangle of the model. Show your work.

4 Write the measurements you found for each triangle on the drawings you made in item 2.

Build the Model

Use the measurements you just made to draw, cut out, and assemble a scale model of your road.

1 Use a protractor and ruler to make accurate scale drawings of your four triangles. Use heavy paper, or cardboard if it's available.

2 Cut out the triangles and tape them together to form the model.

3 Do the sides fit together accurately? If they don't, go back and check your calculations.

How can you make scale drawings of the triangles for your road model?

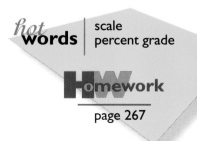

hot **words** | scale percent grade

Homework

page 267

Slopes and Slope-o-meters

Applying Skills

Estimate the slope angle of each line.

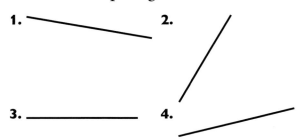

1.
2.
3.
4.

5. a. Use your Slope-o-meter to measure the slope angle of each line in items 1–4. Hold this page against a wall to use the Slope-o-meter.

b. Make a table comparing your estimates in items 1–4 with your Slope-o-meter measurements.

Extending Concepts

6. Explain how you could use your Slope-o-meter to draw a line with a slope angle of 32° on a chalkboard.

7. The three lines below have slope angles of 20°, 35°, and 60°, respectively.

a. Suppose you measure the slope angles of the three lines incorrectly by placing the vertical edge of the Slope-o-meter on the line instead of the horizontal edge, as shown. What do you get for the slope angles of the three lines?

b. For each line, find the sum of the actual slope angle and the slope angle that you measured in item **a.** What do you notice? If you used the incorrect method as described in item **a,** what would you get for a line with slope angle 42°? Explain your answer.

c. If correctly used, does a Slope-o-meter measure slope angles from the horizontal or from the vertical? If used incorrectly as described in item **a,** does it measure slope angles from the horizontal or vertical?

Making Connections

8. The Leaning Tower of Pisa was built in 1174 in Tuscany, Italy. It is 180 ft tall. What would you expect the slope angle of the tower to be if it were not leaning? Use your Slope-o-meter and the illustration shown here to measure the slope angle of the tower.

Working for Scale

Applying Skills

1. a. Estimate the slope angle of this roof.

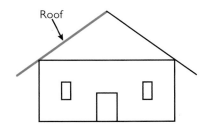

Roof

b. Use your protractor to measure the slope angle. How did your estimate in item **a** compare to the actual slope?

Using a scale of 1 cm = 1 m, find the length of a ramp on a drawing if its actual length is:

2. 48 m   **3.** 13 m   **4.** 7 m

5. Use the scale drawing shown and a ruler to answer these questions.

Scale Drawing
1 cm = 10 m

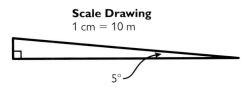

5°

a. What are the height and length of the ramp in the drawing in centimeters?

b. What are the height and length of the actual ramp in meters?

6. Using a scale of 1 cm = 1 m, make an accurate scale drawing of a ramp that has a slope angle of 25° and a length of 14 m.

Extending Concepts

Choose a scale and make an accurate scale drawing using a ruler and a protractor to find the height of each ramp.

7. slope angle 30°, length 7 m

8. slope angle 30°, length 14 m

9. slope angle 60°, length 14 m

10. Use your results in items **7–9** to answer these questions.

a. If two ramps have the same slope but the second ramp is twice as long as the first, do you think that the second ramp must also be twice as high? Explain your thinking.

b. If two ramps have the same length but the slope of the second ramp (in degrees) is twice the slope of the first, do you think that the second ramp must also be twice as high as the first? Explain your thinking.

Writing

11. Answer this letter to Dr. Math.

> Dear Dr. Math:
> What's so special about making a scale drawing? I know the lines all have to be straight but I'm pretty good at drawing straight lines freehand. So why do I need to use a ruler and protractor?
> —"Steady Hand" Luke

Sighting and Angle of Elevation

Applying Skills

1. Use your Slope-o-meter to measure five slopes by sighting. Include at least two angles of elevation and two slanted roofs. For each item, make a sketch showing the angle you measured. For angles of elevation, sketch the invisible "line of sight" with a dotted line.

2. Maria measured the angle of elevation to the top of a redwood tree from four different distances: 10 ft, 25 ft, 48 ft, and 60 ft. She obtained angles of elevation of 84°, 59°, 64°, and 76°, but the angles are not listed in the correct order. Match each distance with the correct angle of elevation.

Extending Concepts

For items 3 and 4, choose a scale and make an accurate scale drawing using a ruler and a protractor to solve the problem.

3. **a.** From a point 12 m from the base of a flagpole, the angle of elevation to the top of the pole is 42°. How tall is the flagpole?

   b. What is the angle of elevation to the top of the flagpole from a point 18 m from the base? Explain how you figured out the answer.

c. Do you think the angle of elevation to the top of the flagpole could ever be equal to 0°? Why or why not? Make a sketch to explain your answer.

4. This drawing of a lake and hill is not drawn to scale. In the drawing, $AB = 600$ m.

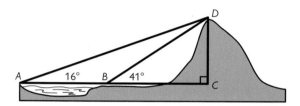

The angles of elevation are: $\angle BAD = 16°$ and $\angle CBD = 41°$. What is the height, CD, of the hill? Explain how you found your answer.

Making Connections

5. Angel Falls in Venezuela is the highest waterfall in South America. From a point 650 ft from its base, the angle of elevation to the top of the falls is 72°. How high is the waterfall? Make an accurate scale drawing and explain how you found your answer. Give your answer to the nearest foot.

Right Triangles

Applying Skills

For each ladder's floor angle, find the wall angle and the slope angle of the ladder.

1. 27°   **2.** 81°   **3.** 49°   **4.** $x°$

Find the measure of the second acute angle in each triangle.

5.
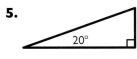
20°

6.

68°

7.
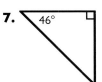
46°

Find the measures of the other angles in a right triangle if the measure of one of the acute angles is:

8. 21°   **9.** 89°   **10.** 5°   **11.** 77°

Extending Concepts

12. An angle in a right triangle is 24° less than one of the other angles. What could the measures of the other angles be? Explain how you solved this problem.

13. The manufacturer of Reach Higher ladders states that on linoleum the maximum stable slope angle is 80° and the minimum is 68°. On carpet the maximum is 85° and the minimum is 30°. Their ladders are 25 ft long.

a. What is the maximum safe wall angle on carpet?

b. What is the minimum safe wall angle on linoleum?

c. At what height will the ladder touch the wall if it stands on carpet with the maximum stable slope angle? with the minimum stable slope angle? Make scale drawings to answer the questions.

Making Connections

14. A **tessellation** is a repeated geometric design that covers a plane without gaps or overlaps. The triangle at right can be used to form the beginning of a tessellation as shown.

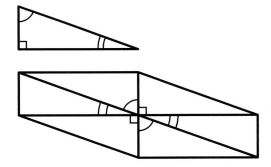

What is the sum of the measures of angles that meet at the center of this pattern? Use what you learned in this lesson about the acute angles of a right triangle to explain how you know that your answer is correct.

Exploring the Pythagorean Theorem

Applying Skills

Write the Pythagorean Theorem equation corresponding to each triangle.

1.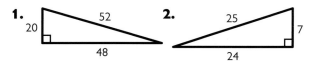

2.

Find the unknown side length in each triangle.

3.

4.

5.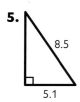

6.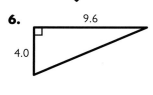

Find the missing value in each set of Pythagorean triples.

7. 10, 24, ?

8. 21, ?, 35

9. ?, 52, 65

10. 15, 36, ?

Extending Concepts

11. a. Verify that the set 20, 48, 52 is a Pythagorean triple.

b. Do you get another Pythagorean triple if you divide each number in the set 20, 48, 52 by 4? by 8?

c. Do you agree with this statement: "If you divide each number in a Pythagorean triple by a constant, you will always get another Pythagorean triple"? If the statement is false, can you modify it to make it true?

d. If you add a constant to each number in a Pythagorean triple will you get another Pythagorean triple? How did you decide?

Writing

12. Answer this letter to Dr. Math.

Dear Dr. Math:

In a right triangle, doesn't $a^2 + b^2 = c^2$? I used this to find the length of the missing side in this triangle.

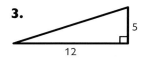

Here is my work: $8^2 + 12^2 = x^2$

So, $x^2 = 64 + 144 = 208$

$\sqrt{208} = 14.42\ldots$

But I know this is wrong because the missing side must be shorter than 12. What did I do wrong?

—Emile

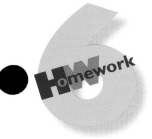

Using the Pythagorean Theorem

Applying Skills

Use the Pythagorean Theorem to find each unknown side length. Give each answer as a decimal rounded to the nearest hundredth.

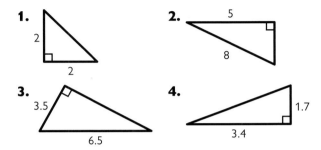

1.

2

2

2.

5

8

3.

3.5

6.5

4.

1.7

3.4

In items 5–7, draw a sketch and use the Pythagorean Theorem to answer the question. Round answers to the nearest hundredth.

5. Find the length of a ramp that covers a horizontal distance of 33 ft and a vertical distance of 7 ft.

6. A 25-ft-long ladder is leaning against a wall of a building. The foot of the ladder is 8 ft out from the bottom of the wall. How high up the wall does the ladder reach?

7. Find the horizontal distance covered by a ramp if the length of the ramp is 45 ft and the vertical distance covered is 10 ft.

Extending Concepts

8. Carefully draw a right triangle having legs 17.3 cm and 19.8 cm long. Measure the hypotenuse. Then use the Pythagorean Theorem to calculate the length of the hypotenuse. Can the Pythagorean Theorem still be true even if your calculated answer and the measured length are different?

9. A steel beam spans a building. The wall on one side of the building is 25 ft tall and the wall on the other side is 32 ft tall. The beam is 38 ft long and extends 2 ft past each wall. How far apart are the walls? Explain how you found your answer.

38 ft

25 ft

32 ft

?

Making Connections

10. The Great Pyramid of Khufu (2680 B.C.) in Egypt is one of the Seven Wonders of the World and is the largest pyramid ever built. It originally measured 756 ft along each side of its base and 482 ft high. Use the Pythagorean Theorem to find the slant height of the original pyramid. Round your answer to the nearest foot.

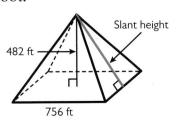

Slant height

482 ft

756 ft

Stairs and Ratios

Applying Skills

Find the slope ratio for each set of stairs.

1. 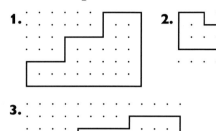 **2.**

3.

Use a ruler (or centimeter dot paper) and a protractor for items 4–6.

4. Order the slopes below from least to greatest.

$$\frac{5}{9} \qquad 35° \qquad \frac{1}{4} \qquad \frac{3}{7} \qquad 20°$$

5. Draw a three-step stairway for each slope ratio.

a. $\frac{5}{4}$ **b.** $\frac{5}{7}$ **c.** $\frac{2}{3}$

6. Use your protractor to measure the slope angle of each stairway in item **5.**

Extending Concepts

7. a. Find the slope ratio for each step in the two-step stairway shown. What is the average of the slope ratios?

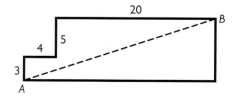

b. What are the total horizontal run and the total vertical rise of the stairway? How did you find them?

c. If the stairway were replaced with a ramp from point *A* to point *B*, what would the slope ratio of the ramp be? Is this the same as the average slope ratio that you calculated in item **a**?

8. If the slope angle corresponding to a slope ratio of $\frac{3}{4}$ is 37°, what slope angle would correspond to a slope ratio of $\frac{4}{3}$? Make a sketch and explain how you found your answer.

Writing

9. Answer this letter to Dr. Math.

> Dear Dr. Math:
>
> My town has a problem. People in wheelchairs can't get into the library because there are 8 steps to the door. The slope ratio of the steps is $\frac{3}{7}$. Architects tried to design a ramp but they said there isn't enough space and that the ramp would end up in the middle of the street. If there's enough space for the stairs, wouldn't there be enough space for a wheelchair access ramp?
> —Left out in Lewiston

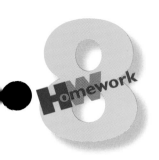

The 12-Inch Tread

Applying Skills

Solve each proportion.

1. $\dfrac{4}{7} = \dfrac{8}{x}$ **2.** $\dfrac{x}{18} = \dfrac{2}{3}$ **3.** $\dfrac{7}{8} = \dfrac{x}{3.5}$

4. $\dfrac{642}{700} = \dfrac{x}{100}$ **5.** $\dfrac{2.4}{3.6} = \dfrac{10}{x}$

For each slope ratio, find the rise that goes with a 12-inch tread.

6. $\dfrac{3}{5}$ **7.** $\dfrac{7}{6}$ **8.** $\dfrac{7}{3}$

For each slope ratio, find the rise that goes with a 10-inch tread.

9. $\dfrac{4}{5}$ **10.** $\dfrac{7}{2}$ **11.** $\dfrac{9}{4}$

Extending Concepts

12. You have been asked to design a stairway with a total rise of 16 ft. You must follow these guidelines:

- The unit rise must be between 6 in. and 8 in.

- unit rise + unit run = between 17 in. and 18 in.

The stairway may have a different number of steps from the one shown.

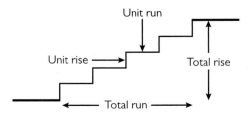

a. What are the maximum and minimum number of steps?

b. What are the maximum and minimum possible values for the total run?

c. How many steps would you recommend?

d. What is the unit rise of your recommended stairway?

e. What value would you recommend for the unit run?

f. What is the total run for your recommended stairway?

Making Connections

13. A **ziggurat** was a type of temple common to the Sumerians, Babylonians, and Assyrians. It was a pyramid-like structure, built in receding steps on a rectangular platform.

Suppose that a ziggurat has the dimensions shown in this cross-sectional view. Assume all steps have the same rise and the same run.

a. What are the rise and run of each step? How did you find them?

b. What is the slope ratio of each step?

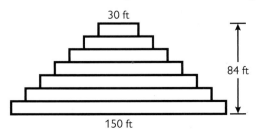

7% Grade Ahead!

Applying Skills

Convert each percent to a ratio or each ratio to a percent.

1. 75%

2. $\frac{2}{5}$

3. $\frac{7}{20}$

4. 55%

For each slope angle, make a scale drawing to find the rise per 200 cm of run. Round your answer to the nearest centimeter. Then complete the table by finding the slope ratio and percent grade.

	Slope Angle	Rise per 200 cm	Slope Ratio	Percent Grade
5.	23°			
6.	47°			
7.	72°			

Find the indicated missing length in each triangle.

8.
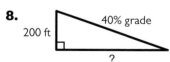
200 ft   40% grade   ?

9.
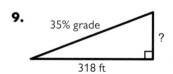
35% grade   ?   318 ft

10. Find the horizontal run of a ramp having a vertical rise of 45 ft and a percent grade of 18%.

11. Find the vertical rise of a ramp having a horizontal run of 71 ft and a percent grade of 32%.

Extending Concepts

12. You are an employee of Ramps Galore and have been asked to design a skateboard ramp with a percent grade of 25%. The ramp must have a vertical rise of 3 ft, and 6 equally spaced vertical braces.

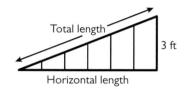
Total length   3 ft   Horizontal length

a. Find the horizontal length of the ramp. How did you figure it out?

b. Use the Pythagorean Theorem to find the total length of the ramp.

c. How far apart should the braces be?

d. Find the height of each brace. Explain how you found your answers.

e. Make a labeled diagram of your ramp.

f. What would the total length of the ramp be if you removed the longest brace?

Making Connections

13. As of 1996, the steepest street in the world is Baldwin Street in Dunedin, New Zealand, with a maximum grade of 79%. On a road with a 79% grade, what would be the horizontal run corresponding to a vertical rise of 200 ft? How far would you have to travel along the road to gain 200 ft in elevation? Make a sketch and explain your reasoning.

The Tangent Ratio

Solve each equation. Round angle measures to the nearest degree and other answers to the nearest hundredth.

1. $\tan 53° = x$　　　　**2.** $\tan x = 0.15838$

3. $\tan x = \dfrac{9}{12}$　　　**4.** $0.6 = \dfrac{10}{x}$

5. $\tan 25° = \dfrac{x}{100}$　　**6.** $\tan 63° = \dfrac{10}{x}$

For each triangle, find all unknown side lengths and angles.

7.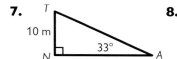

8.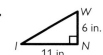

9.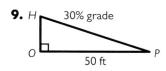

10. Use the Tangent Table to complete the table below.

Slope Angle of Stairs	Tangent Ratio	Percent Grade	Rise for 100 ft Run
34°			
65°			

Extending Concepts

11. A 25-ft ladder is leaning up against the outside wall of a building. The foot of the ladder is 7 ft out from the bottom of the wall. What is the slope angle of the ladder? Make a sketch and explain how you found your answer.

12. You are in a boat in the ocean with your Slope-o-meter.

a. The angle of elevation to the top of a cliff is 35°. You also know the cliff is 1,200 ft straight up above sea level. How far is the boat from the base of the cliff? Make a sketch and calculate the answer using the tangent ratio.

b. What do you notice about the tangent of the angle of elevation as the boat gets very close to the cliff? Why does this make sense?

Making Connections

13. The perpendicular distance from the center of a regular polygon to a side is the **apothem.** The length of each side of the regular hexagon shown is 6. Each interior angle of a regular hexagon measures 120°. Calculate the apothem of the hexagon by using the tangent ratio. Explain how you found your answer.

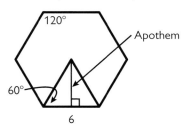

A Mathematical Hill

Applying Skills

Use the figure below to answer the questions.

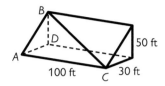

1. For road *AB*, what is the horizontal run? the vertical rise?

2. For road *AB*, find the slope ratio and percent grade and use the Tangent Table to find the slope angle.

3. Make a sketch of triangle *ADC* and label the lengths of the legs. Use the Pythagorean Theorem to find the length of $\overline{CD}$ to the nearest foot.

4. For road *CB*, what is the vertical rise? the horizontal run?

5. For road *CB*, find the slope ratio and percent grade, and use the Tangent Table to find the slope angle.

Extending Concepts

6. a. For road *AB* and for road *CB*, find the horizontal run, vertical rise, and the percent grade.

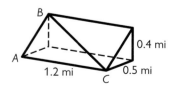

b. What distance would you travel in driving along each road? How did you find your answers?

c. How much further would you drive along road *CB* than along road *AB*? Why might you prefer to drive along road *CB*?

d. Design a road that starts at point *A* and ends at point *B* that is less steep than road *CB*. Make a sketch of your road. Do you think that the distance along your road would be greater or less than the distance along road *CB*?

Writing

7. Answer this letter to Dr. Math.

> Dear Dr. Math:
>
> Last Saturday, my family drove to the top of Mount Washington. The road was so twisty that I got carsick. Why did the road zig-zag?
> I thought that the shortest distance between two points is always a straight line. The road builders would have had much less work to do if they had just built a straight road from the bottom of the mountain to the top.
> —Nauseous in New Hampshire

The Road Project

Applying Skills

For each triangle, find all unknown side lengths and angles.

1.
40°
250 ft

2.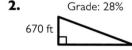
Grade: 28%
670 ft

3.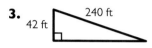
240 ft
42 ft

The picture shows a model of a hill with a slope angle of 32° and a vertical rise of 300 ft. The road *CB* has a grade of 24%.

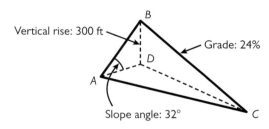

B
Vertical rise: 300 ft
D
Grade: 24%
A
Slope angle: 32°
C

4. Make a sketch of triangle *ADB*. Label any angle measures or side lengths that are known. Use the tangent ratio to find the length of leg *AD*.

5. Make a sketch of triangle *DBC*. Label any known side lengths or angles. Use the 24% grade to find the length of $\overline{DC}$.

Extending Concepts

6. Follow the guidelines to build a model for a road going up a hill. The slope angle of the hill is 36° and its vertical rise is 660 ft. The road *CB* is to have a grade of 27%.

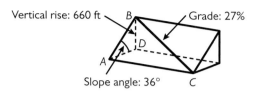

Vertical rise: 660 ft
B
Grade: 27%
D
A
Slope angle: 36°
C

a. Sketch triangles *ADB*, *DBC*, *ADC*, and *ABC*. Label any angle measures, side lengths, or percent grades that are known.

b. Make a list of lengths and angles that need to be found in order to build your model.

c. Find the lengths and angles that you listed in item **b** and write them on the drawings you made in item **a**. Explain how you found them.

d. Use a protractor and ruler to make accurate scale drawings of your four triangles. Explain how you would build the model of the hill.

Making Connections

7. Stonehenge, on Salisbury Plain, England, was built between 1900 B.C. and 1600 B.C. It consists of concentric rings of standing stones. Each stone in the outer circle has approximately the shape of a rectangular prism with height 18 ft, length 7 ft, and width 4 ft. Find the approximate length of a diagonal (*AB*) of one of these stones. Explain how you figured it out.

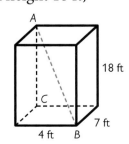

A
18 ft
C
4 ft   *B*   7 ft

Mc Graw Hill Glencoe

This unit of MathScape: Seeing and Thinking Mathematically was developed by the Seeing and Thinking Mathematically project (STM), based at Education Development Center, Inc. (EDC), a non-profit educational research and development organization in Newton, MA. The STM project was supported, in part, by the National Science Foundation Grant No. 9054677. Opinions expressed are those of the authors and not necessarily those of the Foundation.

CREDITS: Photography: Chris Conroy • © Michael L. Marrella: p. 222L.

Send all inquiries to:
Glencoe/McGraw-Hill
8787 Orion Place
Columbus, OH 43240-4027

ISBN: 0-07-866830-1

4 5 6 7 8 9 10 058 06 05